THE STATUS OF ENZYMES AND HORMONES IN THERAPY

BY

G. F. WALKER, M.D., M.R.C.P. (Lond.),

Honorary Physician, Sunderland Royal Infirmary and the Children's Hospital;
Consulting Physician, Sunderland Municipal Hospital;
Honorary Physician, Darlington Memorial Hospital;
Late Medical Registrar in the University of Leeds.

BRISTOL: JOHN WRIGHT & SONS LTD.
LONDON: SIMPKIN MARSHALL LTD.
1935

PRINTED IN ENGLAND BY
JOHN WRIGHT AND SONS LTD.
STONEBRIDGE HOUSE, BRISTOL

THE STATUS OF ENZYMES AND HORMONES IN THERAPY

INTRODUCTION

I SUPPOSE no one will deny that a great deal of confusion—to use no harsher word—exists in therapy with endocrine and similar products. The causes of this are firstly the serious gaps in our knowledge of human physiology, secondly the lamentable lack of co-ordination and control in clinical research, thirdly the docility and credulity with which organic medicaments are handled, and fourthly the everlasting multiplication of trade names.

One of the most serious impediments to the advance of endocrine therapy is the failure on the part of clinicians properly to assess the clinical potency of the remedies they use. Over and over again in the records one can see that 'good results' are recorded from this or that endocrine therapy without adequate recognition of the factors of spontaneous natural amelioration and of therapeutic suggestion. Many maladies for which endocrine preparations are prescribed are capable of natural remissions. And many patients to whom endocrine preparations are given are at a time of life and in a frame of mind when suggestion is most easy and effective.

The purpose of this review is to survey, in as few words as possible, the present status of the numerous remedies of endocrine and organic origin which are to-day so freely used. The landmarks will be the current British Pharmacopœia, the British Pharmaceutical Codex published in September, 1934, and such clinical records selected from the enormous literature as have an honest ring.

He who sets out to use a glandular preparation should ask himself the following questions :—

1. Do I clearly recognize in my patient a hormonic lack ?

2. Do I hold in my hand the truly appropriate remedy for that lack ?

3. Is the preparation stable, active, pure, and uniform ?

4. Can I accurately determine the dosage ?

5. Is the preparation smoothly active when administered by the route I propose ?

6. Are ill-effects impossible ?

If all these questions are answered in the *negative*, it may still be permissible to proceed provided satisfactory answers can be given to :—

7. Is the preparation harmless ?

8. Is the therapy really justifiable either as a simple clinical experiment or under the authority of empiricism ?

If these questions are answered evasively, then the clinician must ask himself whether he can proceed at all.

We can now examine the preparations available. The British Pharmacopœia of 1932 will be designated B.P., and the British Pharmaceutical Codex of September, 1934, will be designated B.P.C.

THERAPEUTIC PREPARATIONS

THYROID

This is taken first in deference to tradition. The pituitary is the master-gland of the whole human body, but the thyroid is the most popular and well-known in general therapeutics. Very large quantities are prescribed annually. A large number of patients take thyroid who do not need it, and a good many more are deprived of it because of faulty diagnosis.

There are two appearances in the B.P. :—

Thyroideum (*syn.* Thyroideum Siccum, Dry Thyroid, Thyroid Extract, Thyroid Gland). This substance is a pale powder and the authorized dose is $\frac{1}{2}$ to 5 gr.

Thyroxinsodium. This may be prepared either from natural sources or by synthesis. It is a pale powder and the dose is $\frac{1}{640}$–$\frac{1}{64}$ gr.

The additional appearance in the B.P.C. is Extractum Thyroidei Liquidum.

The commercial preparations of thyroid are very numerous, and appear mostly under the plain proper name. Thyroid is the commonest constituent of numerous compound organic powders, tablets, and elixirs.

Myxœdema and Cretinism.—The therapeutic use of thyroid is remarkably wide. It is widely and properly prescribed in frank myxœdema, and there is probably no single clinical act more satisfactory to all concerned than the accurate recognition and effective treatment of myxœdema. In most cases the maximum requirement will be less than half the official B.P. dose, and a beginning should always be made with the minimum B.P. dose. In cretinism the results may be equally complete and good. There are normal men and women leading active lives who have taken thyroid since cretinism was diagnosed in infancy. But the prognosis in cretinism needs to be cautious. Everything depends on early recognition. Children who have been left as untreated cretins for several years do not do well under treatment. It is legitimate and proper to prescribe carefully watched and controlled thyroid to a goitrous mother who has already had one cretin child.

Obesity.—Probably the commonest clinical use of thyroid is in the treatment of obesity, and it is possible to lay down some general rules. Firstly, in a condition to be presently described as 'chronic benign hypothyroidism' there is often a puffy or flabby obesity which improves marvellously under thyroid. Secondly, the condition known as adiposis dolorosa often improves under thyroid. Thirdly, many cases of obesity are almost wholly exogenous (i.e., gluttonous) and such derive no benefit whatever from thyroid. Fourthly, attempts to reduce weight with intensive courses of thyroid when there is no hypothyroidism may be distinctly dangerous. The problem, of course, is not so simple as the above four statements; every case of obesity must be judged on its merits. But the four rules embody the safest views on the matter.

'Chronic Benign Hypothyroidism.'—Fortunately for purposes of rapid description this term has been invented. It is a condition almost equally common in the two sexes and most prevalent in the fourth and fifth decades. The syndrome should most certainly be more generally recognized. Many patients go untreated because of inaccurate diagnosis. These patients fall into two classes—namely, placid and plaintive. They feel the cold very much and dislike exertion or change. "Not by any means the man he was" sums up the impression gained by friends. The condition is far short of myxœdema, but consists of a general depression of health marked by a tendency to flabby posture, a depression of intellect and mental acuity, pains in the subcutaneous and fatty tissues round the shoulders and scalp, dryness of the skin, loss of hair, and a liability to catarrh of the mucous membranes. A great deal of abdominal pain is due to this hypothyroidism. Very great benefit accrues from carefully controlled thyroid therapy.

'Myxœdema Heart.'—The condition known as the 'myxœdema heart' may occur without any frank symptoms or signs of myxœdema or hypothyroidism. The condition is not very common, and its recognition depends very largely upon the result of a therapeutic test. It may occur in persons past middle age who present cardiac insufficiency without any obvious common cause. The signs are cardiac enlargement, enfeebled pulse, and slow rate. The electrocardiogram shows low voltage in all complexes and inverted T waves in 1 and 2. The prognosis under treatment is not very good.

Arthritis. — For many years clinicians have recognized and so treated several varieties of arthritis as due to hypothyroidism. Here again the prognosis must be guarded. A certain amount of welcome improvement can nearly always be secured, but it is not safe to expect cure.

Sub-normal Children.—Very large amounts of thyroid are used in the medication of sub-normal or defective children, and thyroid forms the basis of a good many compound tablets similarly used.

Apart from frank cretinism there are many types of hypothyroidism in children, and the following rules will serve as a guide. Firstly, thyroid will bring a subnormal basal metabolism to normal where stability may be expected after the thyroid is discontinued. Secondly, thyroid will promote bodily growth in certain children of deficient stature but normal metabolism. Thirdly, in children, thyroid will not advance a normal basal metabolism. These rules can be applied with prospects of success to the well-known 'debilitated child' in whom hypothyroidism may be at least a factor. The 'debilitated child' is dull and flabby and prone to never-ending vague invalidism. He suffers from worms, constipation, catarrh, adenoids, albuminuria, enuresis, lordosis, chorea, insomnia, and anorexia. He is mentally willing, but slow. No one would suggest that thyroid will cure this condition, but the administration may be the first step towards general improvement.

Gynæcological Conditions.—In gynæcology thyroid is widely and indiscriminately used. The following conditions may be improved with thyroid: amenorrhœa of defective sex development, menorrhagia of climacteric hypothyroidism, and failure of ovulation in hypothyroidism, but that is about as far as there is general agreement. There is, however, hardly any gynæcological condition of which it is impossible to find reports alleging good results from thyroid therapy. Conditions as wide apart as pruritus, eclampsia, and repeated miscarriages are often treated with thyroid, and more clinical acumen would often be an advantage in these and similar procedures. There is especially urgent need of clinical assessment of the place of thyroid in the handling of eclampsia.

Sepsis.—Not long ago thyroid—with or without potassium permanganate—had a vogue in the treatment of boils and sundry infections. It was thought that thyroid exerts a detoxicating and anti-infective action. The original clinical reports are rather more convincing than one would expect, and this form of therapy has possibly not yet had the wide general trial which it may deserve.

Dermatological Conditions.—Thyroid is widely and properly used in dermatology in cases where the dry skin of hypothyroidic states first attracts attention. In myxœdema there is a depression of all the vital activities of the skin. Dermatologists also frequently secure good results with thyroid in the treatment of chilblains.

Miscellaneous Uses.—Great hopes were recently raised that thyroid would act as a preventive of post-operative thrombosis and pulmonary embolism. These hopes have not been fulfilled.

There is now a complete range of evidence that the thyroid promotes red-cell formation, but the therapeutic applications of this are not yet worked out.

Many allergic conditions including migraine, asthma, and certain types of enuresis improve under thyroid.

Thyroid is indeed indicated in *any* condition in which there is depressed metabolism, and hence it has a range of utility extending from anorexia nervosa to ulcers of the leg.

To return to nomenclature and dosage. The B.P. has now made an end of the confusion arising from terms like 'fresh gland', but it seems a pity that the term 'thyroid extract', which is not strictly correct, should be officially retained.

The utmost caution in dosage is required when prolonged medication is proposed. The full B.P. dose given twice or thrice daily for long periods would almost certainly induce severe hyperthyroidism even in out-and-out myxœdema.

The general opinion to-day is that synthetic preparations of thyroxine have no clinical advantages over the natural product, but more may be heard of a substance known as 3–5 Diiodothyronine, which can be prepared in great purity, uniformity, and stability.

OVARIAN HORMONES

For many years tablets, powders, elixirs, and syrups containing, or reputed to contain, ovarian or corpus luteum tissue have been prescribed in large amounts and with singular lack of discrimination in all manner of menstrual, obstetric, and gynæcological disorders. There is probably no field of therapy where a scientific clinical overhaul is more required. It happens that immense progress has recently been made in physiological knowledge, and the result of this is a remarkable increase in proprietary preparations, under new names, containing, or reputed to contain, ovarian hormone.

The B.P. contains no mention of any ovarian substance whatever.

The B.P.C. contains the following :—

Œstrinum.

Corpus Luteum (*syn.* Corpus Luteum Siccum, Desiccated Corpus Luteum).

When, however, one turns to the commercial products the wealth and variety are truly embarass-ing. Considering merely those which are freely advertised one finds substances : (I) For oral use ; (II) For hypodermic use.

I. FOR ORAL USE

The substances for oral use are :—

1. Sundry dried, compressed, or powdered preparations of ovary.
2. Sundry fluid preparations.
3. Sundry preparations containing, or reputed to contain, corpus luteum.
4. Innumerable compound and mixed preparations, some of them with formulæ open to the gravest criticism and scepticism, to say the least.
5. Preparations containing the œstrogenic substance theelol, or the same under another name.

There is room for a comprehensive overhaul of the whole subject of therapy with ovarian tablets, powders, and elixirs. On the one hand, many experienced practitioners, both veterinary and medical, use these preparations with the greatest zeal and persistence, seeing apparent benefit in sundry menopausal disorders and functional uterine disturbances. On the other hand, pharmacologists and many eminent gynæcologists state positively and emphatically that the bulk of these preparations have no direct therapeutic action whatever. They say that good results which are reported are due to suggestion or natural remission or adjuvant measures.

The evidence against the oral use of ovarian substances is piling up very rapidly, but two comments may be made: theelol is perhaps feebly active by the mouth, and its reputed effect is similar to that of œstrin, to be described a little later. Corpus luteum is perhaps feebly active by the mouth and its reputed effect is described in the next section.

II. FOR HYPODERMIC USE

1. Preparations Containing Œstrogenic Substances.—These at present (January, 1935) are perhaps the endocrine preparations most prominently before the profession. The trade names are being multiplied very rapidly. Most of these preparations contain, or are said to contain, the hormone variously known as folliculin or œstrin. This hormone has the following powers :—

It produced œstrus in sexually mature animals. (This is not identical with menstruation in women).

It produces moderate mammary hypertrophy.

It may *depress* ovarian function and inhibit ovulation.

It enhances the rhythmic contractility of the uterus.

It sensitizes the uterus to the action of the posterior oxytocic hormone.

It may have action on secondary gynæcological symptoms, of which more anon.

Clinically the prolonged administration of œstrin brings about: Some enlargement of the uterus with endometrial hyperplasia, a laxity of the pelvis, irregular uterine bleeding, mammary hypertrophy.

The present status of œstrin might therefore be summed up as follows :—

a. The œstrogenic substances have no action—except a possible deleterious action—upon the ovary.

b. With adequate dosage — probably several hundred thousand rodent-units, assisted by appropriate progestin therapy—menstruation can be brought about in a patient presenting amenorrhœa if such amenorrhœa is strictly due to ovarian hypofunction. The question always arises whether menstruation, *per se*, is desirable or advantageous to anyone. There are also doubts whether artificially induced menstruation is worth while.

c. In dosage and manner of administration which has not yet been worked out œstrogenic substance may relieve the subjective symptoms of the menopause, natural or artificial. There is great need for honest and prolonged clinical research on this point. Some clinicians have had very good results in the treatment of menopausal depression, pruritus, arthritis, and eczema.

d. Œstrogenic substance may have a place in hastening delayed puberty, but this is a field of therapy which is almost wholly unexplored. It is likely that œstrin, or some subtance like it, is responsible for the descent of the testes.

e. As a form of substitution therapy œstrogenic substance might conceivably have a place in excessive uterine bleeding. Those who have had most experience of this state that they are most disappointed.

f. As a form of substitution therapy, œstrogenic substance might be used in attempts to regenerate the uterine mucosa and check leucorrhœa.

2. **Corpus Luteum Substance.**—These are preparations containing, or reputed to contain, the active substance from the corpus luteum called corporin or progestin. This hormone has the following known action :—

It sensitizes the uterine mucosa for the safe reception and retention of the fertilized ovum.

It inhibits the normal rhythmic contractility of the uterus.

It prolongs gestation.

It relaxes the pelvic ligaments.

There is a serious lack of clinical reports, even from Germany, respecting prolonged clinical administration of corporin or progestin. We simply do not know what the stuff can do. There is a certain amount of reputable evidence that orally-administered corpus luteum will assist in the safe reception of the fertilized ovum in the uterine mucosa. Hence medication with corporin or progestin, when these are available, might be useful in those cases of sterility where disorder is due to repeated early miscarriage. There is more than a hint in the very voluminous literature that the relief of dysmenorrhœa will ultimately come by means of therapy with progestin.

There remain a few therapeutic procedures which do not fall conveniently into the above grouping :—

1. Ovarian residue by hypodermic injection appears to have a place as a safe uterine stimulant in labour. But claims that this substance is a heaven-sent analgesic have not been sustained. There is a very important piece of work waiting to be done in this connection.

2. Ovarian substance, either by oral or hypodermic route, is said to confer benefit to those suffering from hæmophilia. The published records on this point are curiously vague and contradictory, but one is left with a feeling that there 'may be something in it'. The ovary is the one important organ which hæmophiliacs lack.

To sum up the whole topic of ovarian therapy, it seems that, firstly, we are hindered by stodgy masses of utterly inert preparations ; secondly, we now possess in œstrin and progestin powerful and effective agents ; and thirdly, we have massive problems of unrelieved sufferings and disabilities effecting pretty well the whole of adult female life. There is a gap between the second and the third which no one can contemplate with equanimity or complacency.

PITUITARY HORMONES

The B.P.C. mentions Pituitarium (*syn.* Dry Pituitary Gland, Dry Pituitary, Desiccated Pituitary).

THE ANTERIOR PITUITARY

The B.P. contains no mention.

The B.P.C. mentions, in very reserved and provisional fashion, Pituitary (Anterior Lobe) Extract.

The gap running all through medical practice between physiological knowledge and clinical application is nowhere better illustrated than in a study of anterior pituitary pharmacology. Of this gap much could be said.

Several entirely different hormone-groups have been identified in the anterior pituitary. These include :—

1. The growth-stimulating hormone ('tethelin').
2. The sex-stimulating hormone or hormone group.
3. The thyroid-stimulating hormone.
4. The mammary-stimulating hormone ('prolactin').
5. Hormones regulating sugar metabolism.

Therapeutic application has been most tried in (1) and (2) and hardly at all in (3), (4), and (5).

Available for clinical use the practitioner has a very large array of commercial preparations. These may be classified as follows: (1) For oral administration — Sundry tablets, powders, and syrups. (2) For hypodermic administration — A small number of aqueous extracts and solutions of the anterior lobe (sex-stimulating) hormones and two or three preparations containing hormones resembling the anterior lobe (sex-stimulating) hormones but derived from the urine of pregnant women, or mammalian amniotic fluid or placenta. Some of these preparations are not stable for more than a few weeks.

It is noteworthy that there are no pure preparations of the growth-stimulating hormone, in stable condition and effective dosage, readily available for clinical use. Advances in this connection are, however, imminent. There are a few cases in which alkaline extracts of the anterior pituitary, reputed to be rich in the growth-stimulating hormone, have been used successfully in Simmond's disease or anterior pituitary cachexia.

1. **Preparations for Oral Administration.**—There is good reason to believe that anterior pituitary preparations by the mouth have a definite though feeble ameliorative effect in conditions like Fröhlich's syndrome, the Lorain and Brissaud types of infantile obesity, and in the many ill-defined conditions in which varying degrees of mental and physical retardation, obesity, impaired genital development, and dwarfism are seen.

Results are, of course, difficult to assess, and this is all the more true when anterior pituitary therapy is combined with thyroid, as is often the practice. If the subject of anterior pituitary therapy were tackled firmly with real safeguards as follows, some highly valuable and interesting clinical findings would be at hand. The safeguards would have to be :—

a. The patient must be suffering from a combination of dwarfism, obesity, and sexual retardation attributable to deficiency of his anterior pituitary.

b. The malady must be stationary.

c. The preparation used must be stable and pure.

d. Adjuvant therapy must not be given.

2. **Preparations for Hypodermic Administration.**—The position with regard to sex-stimulating hormones of the anterior pituitary (or hormones resembling these, but actually derived from pregnancy urine) intended for hypodermic use is not at all satisfactory. Apart from possible instability or impurity of the preparations, it has to be confessed that we do not know very much about their dosage, action, or remote effects in the human subject. Quite possibly, when dosage and precise indications are worked out, and when we are sure that ill-effects are impossible, the anterior pituitary hormones by hypodermic medication will have a place in the treatment of conditions such as functional uterine bleeding due to deficient ovarian

activity; sterility due to deficient ovulation; retarded adolescence; deficient formation of corpus luteum; deficient mammary activity; deficient spermatogenesis. But these desirable powers are not yet in our service.

There are, finally, in the literature all manner of accounts of the treatment of the most widely-different conditions by anterior pituitary. Some of these accounts, not quoted herein, merely illustrate man's unique power of self-deception. The position with regard to anterior pituitary therapy is truly lamentable, and the lack of knowledge is all the more to be deplored because it is abundantly clear that the anterior pituitary holds the secret of most of the things which matter in life.

THE POSTERIOR PITUITARY

The B.P. has Extractum Pituitarii Liquidum (*syn.* Liquor Pituitarii, Solution of Pituitary, Pituitary Extract). The dose is given as 2–5 units.

The B.P.C. contains additional mention of Oxytocic Principle of the Pituitary Posterior Lobe; Pressor Principle of the Pituitary Posterior Lobe.

Commercial preparations of posterior pituitary for hypodermic administration, conforming more or less to the B.P., are numerous, well-known, and reliable. There are also wide ranges of tablets, powders, and syrups.

The oxytocic elements and the pressor elements in pituitary extracts can be obtained separately.

The use of pituitary extract is very well established as follows :—

1. As a stimulant to uterine contracture (provided there is no reason to withhold such stimulus).
2. As a circulatory stimulant in all kinds of shock.
3. As a substitution therapy in diabetes insipidus.
4. As an adjuvant to adrenalin in allergic conditions.
5. As a stimulus upon lax or exhausted non-striated muscle. The clinical applications of this in conditions of renal, ureteral, vesical, or alimentary atony are well known.
6. The relief of pain in herpes zoster.
7. To accelerate drainage of the upper urinary tract.

All the above forms of therapy are by hypodermic medication. The only generally accepted form of therapy by other routes is the intranasal insufflatio of dried posterior powder in diabetes insipidus.

Pituitary extract is one of the eight drugs which an experienced practitioner would most zealously retain supposing some dictatorial edict were to limit him to eight. In the relief of pain and shock, in the tiding over of emergencies, and as a genuine pharmacological agent under the above seven principles, pituitary extract has an established place in therapy.

THE THYMUS

No mention of this gland occurs in the B.P.

The B.P.C. contains Thymus (*syn.* Thymus Siccus).

A very large number of commercial preparations are available in both desiccated and aqueous form. They are quietly taken up from time to time for this or that condition and quietly dropped. For some reason thymus is a favourite component of many compound glandular preparations.

The only thymus preparation worth considering is a mixture of thymus and pituitary known as thymophysin. This drug was intended to add speed and safety to parturition and to mitigate the pain in labour. At one time there was an impressive array of support for this. The preparation was deemed of value in many kinds of uterine inertia and also for the induction of labour.

Thymophysin has quite properly been widely used, but the modern view is that its value is no more and no less than the value of the pituitary within it.

THE SUPRARENAL GLANDS

There is no mention in the B.P.

The B.P.C. contains Suprarenalum (*syn.* Dry Suprarenal Gland, Suprarenal Siccum, Desiccated Suprarenal).

THE CORTEX

This substance does not appear in the B.P.

The B.P.C. contains Extractum Suprarenali Corticis (*syn.* Cortin).

Commercial preparations include a small number of dried and powdered preparations, also elixirs and syrups of many kinds. All these are perhaps feebly active when taken by the mouth.

For intravenous or hypodermic administration the pure standardized hormone of the cortex can be obtained commercially under various names.

The function of this substance is unknown. It has remarkable powers as a form of substitution therapy in Addison's disease. In this malady the administration of cortical hormone brings about :—

Relief of nausea and vomiting. Improved appetite.

Increased bodily stamina and gain in weight.

Relief from pain.

Relief of insomnia.

Decrease in pigmentation.

Improved mental condition.

Return of blood-pressure to normal.

Increased resistance to infection.

In view of these remarkable effects *in the presence of suprarenal cortical insufficiency* it is not surprising that the hormone has been used, or proposed, in all manner of conditions of exhaustion, sex disturbance, dermatosis, dementia, muscular weakness, debility, anorexia, and depression.

Except in Addison's disease the merit of therapy

with cortical hormone is not yet well proven. There is obviously an extremely wide field for clinical research. Synthetic cortin for oral administration may be available shortly, but cortin of natural origin has been disappointing on the whole when administered by the mouth.

At one time great hopes were raised that cortical extracts would have an anti-cancerous action. No good came of this.

THE MEDULLA

This appears in the B.P. as Adrenalina (*syn.* Adrenalinum, Adrenalin, Epinephrine). The substance is described as a stable pale powder and the dose given is $\frac{1}{600}$–$\frac{1}{120}$ gr.

A second appearance in the B.P. is Liquor Adrenalinæ Hydrochloridi (*syn.* Hydrochloric Solution of Adrenalin). The dose is 2–8 min. subcutaneously.

The B.P.C. contains also :—

- Insufflation Adrenalinæ.
- Nebula Adrenalinæ Aromatica.
- Nebula Adrenalinæ et Cocainæ.
- Nebula Adrenalinæ et Ephedrinæ.
- Nebula Adrenalinæ et Ephedrinæ Oleosa.
- Suppositorium Adrenalinæ.
- Suppositorium Adrenalinæ et Cocainæ.
- Unguentum Adrenalinæ.
- Unguentum Adrenalinæ et Amylocainæ Compositum.
- Unguentum Adrenalinæ et Cocainæ.

Most of the standardized commercial preparations conform pretty closely to the B.P. Liquor. Effective amounts of adrenalin are used in large numbers of ointments, sprays, snuffs, and suppositories. Many dried and powdered preparations of the medulla intended for oral administration must be either inert or toxic. Nevertheless a few reports are published from time to time indicating that oral administration of dried whole-gland tablets have a place in the treatment of backward children.

The action of adrenalin is very well known. In summary and in principle it may be said that its action is comparable to, and parallel with, stimulation of the sympathetic nervous system. Accordingly it constricts vessels innervated by the sympathetic and increases blood-pressure. It stimulates the cerebral vagus centre and causes a slowing of the heart-beat. It constricts the vessels in oozing surfaces or where it is desirable to retard absorption of another drug. It may be used as a direct stimulant to the muscles of the heart. It is widely used in the relief of maladies like migraine, serum shock, anaphylaxis, and notably asthma. It has a place in the treatment of heart-block.

THE PARATHYROID

There is no appearance of this gland in the B.P. The B.P.C. contains :—

Extractum Parathyroidei.
Parathyroideum.

Tabellæ Parathyroidei et Calcii et Sodii Lactatis.
Tabellæ Parathyroidei et Calcii Lactatis.

Commercial preparations include numerous powders and tablets, many of which are little more than inert scraps of slaughter-house offal.

Some preparations intended for oral administration have a feeble activity, but clinical attention is being given to pure, standardized, and stable preparations of the hormone—originally called 'parathormone'—which are now available for hypodermic administration. Parathormone raises the amount of serum-calcium by withdrawing calcium from the first available source. This powerful preparation needs the most careful handling. Overdosage can produce alarming and disastrous general symptoms the exact nature of which has not been worked out. Prolonged administration of parathormone is not always possible because, curiously enough, patients acquire an 'immunity' to it.

The action of parathormone may be stated in summary as an elevating effect upon the calcium content of the blood, more especially when this is depressed by any pathological condition. Calcium in assimilable form should be given along with parathormone.

The rational use of parathormone is thus *at present* remarkably limited. It is of great value in any condition of hypo-parathyroidism such as tetania parathyreopriva. In infantile tetany and sundry manifestations of spasmophilia it may be a valuable temporary expedient pending a rectification of the underlying cause. And in the future when exact doses and indications have been worked out, it may be of value in those conditions where an increase of serum-calcium is desirable. These conditions may be chronic suppuration, hæmorrhage from any surface, chorea, osteitis deformans, tuberculosis, chilblains, chronic ulceration, and cardiac failure.

Parathormone is said to have the power of evicting lead from the bones in cases of chronic plumbism.

The preparations of parathyroid intended for oral administration are widely used. Otosclerosis and rhinorrhœa are frequently treated with reports of good results which cannot be ignored. The same remarks apply to chorea. But there is room for an overhaul of these forms of therapy in which parathyroid tablets are used vaguely and without clear indications. In many cases, no doubt, the dosage is far too small. Clinical results are often widely different from, and much better than, theoretical considerations. There is clearly a discrepancy somewhere.

THE TESTIS

The B.P. has no mention, nor has the B.P.C.

Innumerable proprietary preparations of all kinds contain testicular tissue or extracts. Most of these—if not all—represent the triumph of hope over experience.

The male hormone of the testis has been chemically isolated and identified. It is to be called 'proviron', but it is not available on any scale for clinical application. When it becomes available it will need the most rigid scrutiny.

THE SPLEEN

The B.P. has no mention of this organ, nor has the B.P.C.

The spleen is not especially favoured by the manufacturing firms.

Clinical work on therapy with splenic substances and extracts is at present very scanty and uncertain. There is a curiously insistent note in a few continental and transatlantic journals that splenic tissue is of value in tuberculosis. Wheeldon's article on this subject—a very convincing one—is well worth careful scrutiny for several reasons (*see* p. 46).

As to the hæmatopoietic action, the insulin-adjuvant action, the anti-allergic action, and the anti-infective action, the most that can be said is that these fields have not yet been fully explored. Someone may find a nugget some day.

MUCIN

There is no appearance of this substance in the B.P. or in the B.P.C.

A few commercial preparations are on the

market, and probably mucin is an adventitious component of many mixed-gland tablets and organic powders.

The substance is well spoken of by those who administer it as having a soothing and demulcent action on the gastric and duodenal mucosa. Those who take it deny that it is palatable. There is probably a wide utility in the treatment of gastric, duodenal, and jejunal ulceration, but no one has yet written a convincing large-scale report.

CONNECTIVE TISSUE

Neither the B.P. nor the B.P.C. has any mention, and proprietary preparations are not generally available.

A recent suggestion that connective-tissue extracts might be of value in cancer received far more attention in America than in this country.

BRAIN TISSUE

The B.P. contains no mention, nor does the B.P.C.

Many proprietary preparations are still on the market.

The only observation worthy of inclusion in this report is that ox brain may exert a beneficial effect upon the neurological phenomena of pernicious anæmia.

MUSCLE TISSUE

Neither the B.P. nor the B.P.C. has any mention.

A very large number of preparations reputed to contain muscle extract or tissue are on the market, and it would seem that they have a wide popularity. Both oral and intramuscular preparations have a wide usage.

A certain amount of cautious optimism has been expressed about these substances in the treatment of angina pectoris, hypertension, intermittent claudication, thrombo-angiitis obliterans, and Raynaud's disease. Except in one place the matter has not yet been the subject of any sustained clinical research.

THE LIVER

The B.P. contains :—

Extractum Hepatis Liquidum.

One fluid ounce contains the equivalent of 8 ounces of fresh liver. Dose : 1 fluid ounce.

Extractum Hepatis Siccum.

Dose : The quantity equivalent to half a pound of fresh liver.

The B.P.C. contains nothing additional.

Commercial preparations include large numbers of fluid extracts, powders, tablets, syrups, and ampoules. Straightforward extracts corresponding to the B.P. are perhaps most widely used. Many extracts have large doses of iron included in them. Preparations for hypodermic and intramuscular therapy are readily available and widely used. Some of the intramuscular preparations enable the clinician to give his patient a sort of reservoir or 'depôt' of anti-anæmic substance by means of weekly injections.

Most of the liver preparations on the market are from mammalian sources, but there is a beginning being made with fish-liver. One preparation of fish-liver for anti-anæmic therapy is advertised as being cheaper than any other form of anti-anæmic therapy.

The hypodermic preparations have an obvious place of value in emergencies and with patients *in extremis*. The literature on this point is, however, very scanty.

Pernicious Anæmia.—Liver substance is perfect 'substitution' therapy for pernicious anæmia (macrocytic hyperchromic anæmia) and effective preventive therapy against subacute combined degeneration of the cord, provided—and provided only—that the following conditions are fulfilled :—

1. The diagnosis is accurate. This seems an absurd platitude to insert in a paper of this kind. But much of the disrepute into which liver therapy fell at one time was due to careless diagnosis and wrong usage.

2. That the patient is studied individually over

a long period, and the dosage individually adjusted and regulated from time to time to ensure a satisfactory blood-count and state. One reason why subacute combined degeneration is now so common is that many patients have been taking liver in amounts sufficient to keep them alive but insufficient to keep their cords intact.

3. That a diet rich in protein and vitamins is taken.

4. That really substantial doses of iron are administered if required.

5. That all deleterious processes such as malnutrition and sepsis are effectively dealt with.

6. That the patient has the power and resources for intelligent and permanent co-operation in the instructions given him.

If these provisions are carried out, no patient with pernicious anæmia need suffer actively from his disease, and no patient with pernicious anæmia should develop subacute combined degeneration.

Miscellaneous Conditions.—Liver has been tried in many conditions besides pernicious anæmia. Some of these have been genuine clinical experiments with negative results. Some have been trials worthy of repetition. A classification might be compiled as follows :—

1. Maladies in which liver therapy is effective—
   a. 'Pernicious anæmia of pregnancy.'
   b. Sprue.
   c. The early mental changes in pernicious anæmia.

2. Maladies in which liver therapy has given grounds for cautious optimism and further trial—

a. Certain types of secondary anæmia, especially where there has been blood loss.
b. Certain types of cœliac disease.
c. Pink disease.
d. Disseminated sclerosis.
e. Anæmia of metazoan infestation.
f. Dermatitis following metallic poisoning.
g. X-ray sickness.
h. Certain types of allergy.

3. Maladies in which liver therapy has been of doubtful value or even a complete disappointment—

a. Diabetes mellitus.
b. Hypertension.
c. Rheumatoid arthritis.
d. Hæmophilia.
e. Purpura.

NON-MAMMALIAN LIVER

There is no mention in either the B.P. or the B.P.C.

One or two commercial preparations containing active anti-anæmic principles from fish-liver are now available. The makers claim advantages in the way of cheapness.

STOMACH PREPARATIONS

(Anti-anæmic)

The B.P. makes no mention.

The B.P.C. mentions Ventriculus Desiccatus.

A very considerable number of commercial preparations are available for clinical use. Clinicians who have had most experience are most enthusiastic. It must be noted, however, that many preparations on the market have been found to vary in potency.

Those who have used desiccated or other stomach preparations most freely and most carefully, declare that this substance is a highly efficient anti-anæmic substance ('hæmopoietin' has been the name proposed) and also a highly safe preventive of subacute combined degeneration. It is also claimed by responsible observers that desiccated stomach is cheap and convenient, but these claims are not universally admitted.

PEPSIN

The B.P. recognizes Pepsinum as containing the proteolytic enzymes of the gastric juice of animals and suggests a dose of 5–10 gr.

The B.P.C. has also :—

Elixir Pepsini.
Glycerinum Pepsini.
Glycerinum Pepsini Fortius.
Liquor Euonymini et Pepsini.
Liquor Pepticus.
Mistura Bismuthi Composita Acida cum Pepsino.
Mistura Bismuthi Composita cum Pepsino.
Mistura Bismuthi Composita cum Pepsino et Morphina.

Pulvis Pepsini Compositus.
Vinum Pepsini.

Pepsin is very widely prescribed either alone or as a constituent of tablets or mixtures designed to aid digestion.

These preparations have a wide clinical popularity, but pharmacologists are strangely sceptical as to their value.

PEPTONE

The B.P. makes no mention.

The B.P.C. mentions:—

Peptonum.
Injectio Peptoni.
Liquor Ferri Peptonatis.
Liquor Ferri Peptonatis cum Mangano.
Suppositorium Nutriens.

This substance has been widely used for the treatment of allergic diseases such as asthma, migraine, and hay-fever, and those who employ it most get good results. The most recent method is the intramuscular injection of a 5 per cent aqueous solution. There is room for more clinical reports on this form of therapy.

PANCREATIC PREPARATIONS

The B.P. recognizes:—

Insulinum. Dose 5–100 units.
Pancreatinum. This is a colourless powder containing the enzymes trypsin, amylase, and lipase.

The B.P.C. also mentions:—

Glycerinum Pancreatini.
Liquor Pancreatini.
Mistura Bismuthi et Pancreatini.
Pulvis Pancreatini Compositus.
Pulvis Pepsini Compositus.
Tabellæ Pancreatini.

1. Insulin.—Discussion on the established usage of insulin in diabetes mellitus would be superfluous in a paper of this kind. Time has not yet solved the problem of the fate of the diabetic child, nor can it be yet finally decided that insulin improves endogenous insulin secretion.

There is no oral substitute for insulin in spite of repeated attempts and enthusiastic commercial claims.

Insulin is probably still employed with insufficient resolution and patience in those oppressed by the complications of diabetes. More care and skill would justify more optimism in the treatment of diabetics with phthisis.

Insulin has been widely employed in non-diabetic conditions with what are, on the whole, unsatisfactory results. Many of the 'successful' series of cases lack the value of a later critical follow-up. It appears, however, to have a place in the treatment of cardiac failure, morphinism, states of cachexia or malnutrition, cirrhosis of the liver, and arthritis.

2. Pancreatinum.—This substance is widely prescribed in various forms to aid intestinal

digestion, but pharmacologists are again sceptical, more especially if the vehicle of administration is not proof against the gastric juice. Probably the best form of administration of pancreatinum is in the form of liquor pancreatini about three hours after a meal. Another good method is the use of tablets or capsules which are not digested in the stomach.

The substance has a wide use and undoubted utility for the artificial pre-digestion of milk.

3. Pancreatic Extracts are widely employed in combination with other organic preparations, and some of these tablets have attained to wide clinical popularity in the treatment of hypertensive states. It is suggested that the pancreas has some substance or hormone which checks a tendency towards arterial hypertension. There is a fine opportunity for clinical research here.

DUODENAL SUBSTANCE

There is no mention in the B.P.

The B.P.C. mentions Duodeni Membranum (*syn.* Pulvis Duodenalis, Duodenal Powder).

With important reservations, the B.P.C. also mentions Liquor Duodenalis and Secretinum.

Duodenal substance contains secretin, enterokinase, erepsin, invertase, lactase, and maltase. It is probably chiefly of value in virtue of the enterokinase and to a lesser exent in virtue of the enzymes concerned in the digestion of proteins and sugars.

On the whole, duodenal substance would appear to be worthy of more clinical attention and usage. Apart from the enzyme activity described above, it is more than probable that the duodenum contains an anti-anæmic substance.

THE PLACENTA

There is no mention in the B.P. or in the B.P.C.

A number of commercial preparations exist and placental substance goes towards the making of a good many 'compound' tablets. Some of these are said to combat the zymotic infections, and one reads the accounts of this kind of therapy with complete bewilderment.

The actual hormones lodged in the placenta are œstrin, emmenin (which is probably identical with œstrin), and a hormone resembling the anterior pituitary.

These substances have clinical value as described under previous headings. The use of the placenta as a source of gonadotropic and œstrogenic substances will probably be expanded in the future. Random clinical use at present is much to be deprecated.

BILE

The B.P. mentions Extractum Fellis Bovini (*syn.* Fel Bovinum Purificatum; Purified Ox Bile). Dose 5–15 gr.

The B.P.C. mentions Fel Bovinum, of which the B.P. product is a derivative.

The clinical utilities of bile are chiefly as an aid in the absorption of fats, as an aperient, and as a cholagogue. The popularity of bile as a therapeutic agent is waxing again after lengthy waning.

Bile salts have been used successfully in the treatment of migraine and to combat a tendency to hæmorrhage.

RENAL SUBSTANCE

There is no mention in the B.P. or in the B.P.C.

A tablet purporting to be renal tissue is much advertised and much employed as a diuretic.

The clinical action and value has not been comprehensively worked out.

PINEAL, MAMMARY, PROSTATE, LYMPHATIC SUBSTANCE

There is no mention in either the B.P. or the B.P.C., and one can search the *Index Medicus* from end to end without finding any convincing signs of clinical utility. Yet nearly all the manufacturing chemists market these extracts in some form or other. Somebody must be prescribing them. It should be noted that prostatic substance may be toxic.

BONE-MARROW

The B.P. has no mention, but the B.P.C. mentions Medulla Rubra and a number of preparations such as Extractum Malti Liquidum et Medullæ Rubra.

This substance is quite rightly used in a number of valuable tonic fluids and foods. The exact pharmacology has never been worked out.

HÆMOGLOBIN

No mention in the B.P.

The B.P.C. recognizes Hæmoglobinum, but extends very timid approval. A compound elixir and a malt extract with hæmoglobin are also mentioned.

The value of the copper in natural hæmoglobin has never been worked out.

SUMMARY

The final decision on the value of an organic or endocrine preparation rests with the clinician. Endocrine therapy seems to be doubly cursed with an indiscriminate enthusiasm in some clinicians and a sour scepticism in others. One has a feeling that a good many endocrine preparations have never had a fair trial with proper watch and record of their actions. Along with this feeling is the uncomfortable awareness that much ill-health and suffering could be mitigated by a better general knowledge of the powerful and natural agents whose present unsatisfactory status I have endeavoured to estimate.

BIBLIOGRAPHY

THYROID

General—

Gardiner-Hill, *Lancet*, 1933, ii, 937.
Gardiner-Hill, *Practitioner*, 1931, cxxvii, 627.
Lisser, *Internat. Clinics*, 1933, iv, 66.

In Children—

Topper and Cohen, *Amer. Jour. Dis. Child.*, 1928, xxxv. 205.
Barbour, *Arch. of Pœdiatrics*, 1933, l, 851.
Topper, *Amer. Jour. Dis. Child.*, 1931, xli, 1289.
Rolleston, *Lancet*, 1934, ii, 372.
Gardiner-Hill and Smith, *Quart. Jour. Med.*, 1926, xviii, 309.

Chronic Benign Hypothyroidism—

McKean, *Mich. State Med. Jour.*, 1929, xxviii, 128.
Barlow, *Practitioner*, 1931, cxxvii, 648.
Warfield, *Jour. Amer. Med. Assoc.*, 1930, xcv, 1076.
Hinton, *Jour. Amer. Med. Assoc.*, 1932, xcviii, 1702.

Obesity—

Lyon and Dunlop, *Quart. Jour. Med.*, 1932, i, 331.

Dermatology—

Mumford, *Lancet*, 1927, ii, 863.

Sepsis—

Ball, *Practitioner*, 1933, cxxx, 101.
Nott, *Brit. Med. Jour.*, 1928, i, 94.

Pulmonary Embolism—

Fraser, *Brit. Med. Jour.*, 1932, i, 659.

Arthritis—

Swain, *Jour. Amer. Med. Assoc.*, 1929, xciii, 259.

Cardiac—

Aymen and others, *Jour. Amer. Med. Assoc.*, 1933, xcviii, 1721.

Diiodothyronine—

Anderson and others, *Lancet*, 1933, ii, 1081.

Anorexia Nervosa—

Berkman, *Amer. Jour. Med. Sci.*, 1930, clxxx, 411.

OVARIAN HORMONES

GENERAL—

Novak, *Internat. Clinics*, 1933, iv, 179.

General Report, *Jour. Amer. Med. Assoc.*, 1933, c, 1331.

Finkel, *New Eng. Jour. Med.*, 1933, ccix, 477.

Parkes, *Lancet*, 1934, i, 557.

THERAPY (i) SCEPTICAL—

Frank, R. T., *Jour. Amer. Med. Assoc.*, 1931, xcvii, 1852.

Editorial, *Lancet*, 1932, ii, 525.

Frank, *Jour. Amer. Med. Assoc.*, 1934, ciii, 393.

THERAPY (ii) ENTHUSIASTIC—

Schoeller and others, *Amer. Jour. Med. Sci.*, 1931, clxxxii, 326.

Kennedy, *Brit. Med. Jour.*, 1932, i, 746.

Sandell, *Med. Press and Circ.*, 1934, clxxxix, 191.

THERAPY (iii) CRITICAL AND CONSTRUCTIVE—

Editorial, *Lancet*, 1934, ii, 1236.

Dodds, *Lancet*, 1934, ii, 1318.

Dodds, *B.M.J.*, 1934, ii, 1187.

Kaufmann and others, *Proc. Roy. Soc. Med.*, 1934, Feb. 13.

Dodds, *Proc. Roy. Soc. Med.*, 1932, Jan. 16.

Johnstone and others, *Lancet*, 1932, ii, 509.

Novak, *Brit. Med. Jour.*, 1933, ii, 553.

Johnstone, *Brit. Med. Jour.*, 1933, ii, 557.

Haultain. *Practitioner*, 1933, cxxx, 534.

Editorial, *Lancet*, 1934, i, 469.

Loeser, *Jour. Obst. and Gynæcol. Brit. Emp.*, 1934, xli, 86.

Marshall, *Brit. Med. Jour.*, 1932, ii, 232.

Shea, *Brit. Med. Jour.*, 1933, i, 1053.

Thomson, *Lancet*, 1932, i, 229.

Gardiner-Hill and Smith, *Lancet*, 1931, i, 464.

HÆMOPHILIA—

Birch, *Jour. Amer. Med. Assoc.*, 1932, xcix, 1566.

Freeman, *Irish Jour. Med. Sci.*, 1933, lxxxv, 122.

Foord and Dysart, *Jour. Amer. Med. Assoc.*, 1932, xcviii, 1444.

Stetson and others, *Jour. Amer. Med. Assoc.*, 1934, cii, 1122. (Sceptical.)

Brem and Leopold, *Jour. Amer. Med. Assoc.*, 1934, cii, 200. (Sceptical.)

OVARIAN RESIDUE—

Rodway, *Lancet*, 1933, i, 692.

Cressy, *Brit. Med. Jour.*, 1930, ii, 711.

PITUITARY HORMONES

Anterior Pituitary :—

GENERAL—

Editorial, *Lancet*, 1934, ii, 490.
Editorial, *Brit. Med. Jour.*, 1934, i, 628.
Editorial, *Brit. Med. Jour.*, 1934, i, 111.
Collip, Anderson, and Thomson, *Lancet*, 1933, ii, 347.
Anselmino and others, Abstr. *Brit. Med. Jour.*, 1933, Dec. 9.
Lambie and Wiesner, *Edin. Med. Jour.*, 1931, xxxviii, 605.

THERAPY—

Gardiner-Hill and Smith, *Quart. Jour. Med.*, 1926, xviii, 309.
Brosius and Schaffer, *Jour. Amer. Med. Assoc.*, 1933, ci, 1227.
Calder, *Bull. Johns Hopkins Hospital*, 1932, l, 87.
Campbell, *Canad. Med. Assoc. Jour.*, 1932, xxvii, 347.
Tutcher, *Internat. Clinics*, 1933, iv, 114.
Mayes, *Med. Jour. Australia*, 1933, ii, 677.

Posterior Pituitary :—

Tutcher, *Internat. Clinics*, 1933, iv, 114.
Editorial, *Lancet*, 1934, ii, 490.
Smith, *Jour. Amer. Med. Assoc.*, 1934, cii, 660.
Darley and Draper, *Jour. Amer. Med. Assoc.*, 1934, cii, 667.

THE THYMUS

Greenhill, *Jour. Amer. Med. Assoc.*, 1932, xcviii, 1260.
Roques and MacLeod, *Jour. Obst. and Gynæcol. Brit. Emp.*, 1932, xxxix, 320.

THE SUPRARENAL GLANDS

The Cortex :—

Swingle and Pfiffner, *Medicine*, 1932, xi, 372.
Hartman, *New Eng. Jour. Med.*, 1933, ccix, 480.
Editorial, *Brit. Med. Jour.*, 1934, i, 363.
Editorial, *Lancet*, 1934, i, 1241.
Editorial, *Lancct*, 1933, xi, 195.
Simpson, *Brit. Med. Jour.*, 1932, ii, 625.
Editorial, *Brit. Med. Jour.*, 1933, i, 1013.
Freeman, *Endocrinology*, 1933, xvii, 677.
Rowntree, *Internat. Clinics*, 1933, iv, 155.
Rogoff and Stewart, *Jour. Amer. Med. Assoc.*, 1929, xcii, 1569.
Mendelson, *Jour. Amer. Med. Assoc.*, 1934, cii, 604.

The Medulla:—

Barbour, *Arch. of Pediatrics*, 1933, l, 851.
Gilchrist, *Quart. Jour. Med.*, 1933, xxvi, 499.
Hunt, *Lancet*, 1933, ii, 279.
Packard and Wechsler, *Amer. Jour. Med. Sci.*, 1933, clxxxvi, 66.

THE PARATHYROID

Thomson and Collip, *Internat. Clinics*, 1933, iv, 102.
Editorial (quoting Hunter and Aub), *Lancet*, 1934, ii, 494.
Snapper, *Lancet*, 1934, i, 728.
Editorial (quoting Marx), *Med. Press and Circ.*, 1934, clxxxviii, 434.
Editorial, *Lancet*, 1933, ii, 1267.
Colt and Lyall, *Brit. Med. Jour.*, 1933, ii, 10.
Stephens, *Med. Press and Circ.*, 1932, clxxxvi, 417.
Watson-Williams, *Lancet*, 1929, i, 814.
Haskell, *Amer. Jour. Med. Sci.*, 1931, clxxxi, 180.

THE TESTIS

GENERAL—

Editorial, *Jour. Amer. Med. Assoc.*, 1932, xcviii, 738.

CRITICAL AND CONSTRUCTIVE—

Walker, *Practitioner*, 1933, cxxx, 534.
Parkes, *Lancet*, 1934, i, 557.
Schoeller, *Lancet*, 1933, i, 38.

THE SPLEEN

Wheeldon, *Jour. Bone and Joint Surg.*, 1933, xv, 337.
Editorial, *Brit. Med. Jour.*, 1934, ii, 477.

MUCIN

Editorial, *Jour. Amer. Med. Assoc.*, 1934, cii, 767.
Stuart Low, *Brit. Med. Jour.*, 1931, ii, 124.

CONNECTIVE TISSUE

Baker, *Lancet*, 1933, ii, 643.

BRAIN TISSUE

Ungley, *Lancet*, 1932, i, 227.

MUSCLE TISSUE

Editorial, *Brit. Med. Jour.*, 1931, i, 533.
Hartfall, *Univ. Leeds Medical Society Magazine*, 1934, May. (Sceptical.)

THE LIVER

GENERAL—
Editorial, *Lancet*, 1934, ii, 1290.
Royal Society of Medicine, 1934, Nov. 29 (Lecture by G. R. Minot).
Discussion, Royal Society of Medicine, 1933, Jan. 10.
Discussion, Royal Society of Medicine, 1931, March 24.
Dyke and Harvey, *Lancet*, 1933, ii, 59.
Schultzer, Abstr. *Brit. Med. Jour.*, 1934, June 2.
Murphy, *Amer. Jour. Med. Sci.*, 1933, clxxxvi, 361.
Bethel and Goldhammer, *Amer. Jour. Med. Sci.*, 1933, clxxxvi, 480.
Strauss and Castle, *Lancet*, 1932, ii, 111.
Davidson, *Brit. Med. Jour.*, 1932, i, 269.
Rogers and Cooke, *Brit. Med. Jour.*, 1932, i, 272.
West-Watson, Billig, and Young, *Brit. Med. Jour.*, 1932, i, 273.

SUBACUTE COMBINED DEGENERATION—
Greenfield and O'Flynn, *Lancet*, 1933, ii, 62.
Witts, *Brit. Med. Jour.*, 1934, i, 1407.
Discussion, Royal Society of Medicine, 1933, Dec. 21.

PERNICIOUS ANÆMIA OF PREGNANCY—
Strauss, *Jour. Amer. Med. Assoc.*, 1934, cii, 281.
Strauss and Castle, *Lancet*, 1932, i, 1198.

DISSEMINATED SCLEROSIS—
Gowlland, *Med. Press and Circ.*, 1934, clxxxix, 81.

MENTAL CHANGES—
McAlpine, *Lancet*, 1929, ii, 641.
Atkin, *Lancet*, 1932, ii, 569.

X-RAY SICKNESS—
Webster, *Brit. Med. Jour.*, 1934, i, 15.

SPRUE—
Castle and Rhoads, *Lancet*, 1932, i, 1198.
Bassler and Lutz, *Jour. Amer. Med. Assoc.*, 1930, xcv, 1961.

NON-MAMMALIAN LIVER—
Davidson, *Brit. Med. Jour.*, 1932, ii, 347.

STOMACH PREPARATIONS

Wilkinson, *Lancet*, 1932, i, 719.
Wilkinson, *Proc. Roy. Soc. Med.*, 1933, March 14.
Wilkinson, *Practitioner*, 1933, cxxxi, 409.

Editorial, *Brit. Med. Jour.*, 1932, i, 1179.
Wilkinson, *Brit. Med. Jour.*, 1932 i, 325.
Ferguson and others, *Proc. Roy. Soc. Med.*, 1933, Dec. 21.

PEPTONE

Bray, *Practitioner*, 1933, cxxx, 545.
Auld, *Lancet*, 1932, ii, 67.
Auld, *Proc. Roy. Soc. Med.*, 1934, March 13.

PANCREATIC PREPARATIONS

Insulin :—

Wilcox, *Clinical Jour.*, 1931, lx, 493.
Thomson, *Brit. Med. Jour.*, 1932, i, 322.
Blotner, *Jour. Amer. Med. Assoc.*, 1933, c, 88.
Smith, *Brit. Med. Jour.*, 1933, i, 693.
Allen, *Jour. Amer. Med. Assoc.*, 1933, ci, 1797.
Editorial, *Brit. Med. Jour.*, 1934, ii, 519.
Campbell, *Internat. Clinics*, 1933, iv, 142.
Leyton, *Lancet*, 1933, ii, 120.

Pancreatic Extracts :—

Nuzum and Elliot, *Arch. of Internal Med.*, 1932, xlix, 1007.
Villaret, *Presse méd.*, 1929, xxxvii, 1685.

THE PLACENTA

Editorial, *Lancet*, 1933, ii, 1491.
Collip, *Jour. Amer. Med. Assoc.*, 1933, ci, 1553.
Chu, *Amer. Jour. Dis. Child.*, 1932, xliv, 1712.
McKhann, *Amer. Jour. Dis. Child.*, 1933, xlv. 475.

BILE

Abstr. *Brit. Med. Jour.*, 1934, Aug. 4.
Hunt, *Lancet*, 1933, ii, 279.

RENAL SUBSTANCE

Dejust-defiol et Rome, *Presse méd.*, 1932, xiii, 238.

Printed in Dunstable, United Kingdom

91221478R00030